Playing with Wind

Elizabeth Austen

✹ Smithsonian

© 2019 Smithsonian Institution. The name "Smithsonian" and the Smithsonian logo are registered trademarks owned by the Smithsonian Institution.

Blow, wind, blow!

Fly, kite, fly!

Stop, wind, stop!

Drop, kite, drop!

Blow, wind, blow!

Dip, kite, dip!

Spin, kite, spin!

Fly, kite, fly!

STEAM CHALLENGE

The Problem

It is your job to test wind direction. How can you tell where the wind is blowing?

The Goals

- Make a device to see which way the wind blows.
- It can be made from any materials you like.
- It should blow easily in the wind.

Research and Brainstorm
Learn about wind.

Design and Build
Draw your plan. Build your device!

Test and Improve
Use the device to test wind direction. Then, try to make it better.

Reflect and Share
What did you learn?

Consultants

Amy Zoque
STEM Coordinator and Instructional Coach
Vineyard STEM School
Ontario Montclair District

Siobhan Simmons
Marblehead Elementary
Capistrano Unified School District

Publishing Credits

Rachelle Cracchiolo, M.S.Ed., *Publisher*
Conni Medina, M.A.Ed., *Editor in Chief*
Diana Kenney, M.A.Ed., NBCT, *Series Developer*
Emily R. Smith, M.A.Ed., *Content Director*
Véronique Bos, *Creative Director*
Robin Erickson, *Art Director*
Stephanie Bernard, *Associate Editor*
Mindy Duits, *Senior Graphic Designer*
Smithsonian Science Education Center

Image Credits: all images from iStock and/or Shutterstock.

Library of Congress Cataloging-in-Publication Data

Names: Austen, Elizabeth (Elizabeth Charlotte), author.
Title: Playing with wind / Elizabeth Austen.
Description: Huntington Beach, CA : Teacher Created Materials, [2020] | "Smithsonian Institution"--Copyright statement. | Includes index. | Audience: Age 5. | Audience: K to grade 3. |
Identifiers: LCCN 2018055269 (print) | LCCN 2018060653 (ebook) | ISBN 9781425859886 (eBook) | ISBN 9781493866434 | ISBN 9781493866434¬(pbk.) |
  ISBN 1493866435¬(pbk.)
Subjects: LCSH: Winds--Juvenile literature. | Weather--Juvenile literature.
Classification: LCC QC931.4 (ebook) | LCC QC931.4 .A97 2020 (print) | DDC 551.51/8--dc23
LC record available at https://lccn.loc.gov/2018055269

Smithsonian

© 2019 Smithsonian Institution. The name "Smithsonian" and the Smithsonian logo are registered trademarks owned by the Smithsonian Institution.

Teacher Created Materials

5301 Oceanus Drive
Huntington Beach, CA 92649-1030
www.tcmpub.com
ISBN 978-1-4938-6643-4
© 2019 Teacher Created Materials, Inc.
Printed in Malaysia
Thumbprints.21248